Cattle Journal

Owner: ______________________

Dates: ______________________

Herd: ______________________

Beech Cottage Press
Lovingston, Virginia

Published June 21st, 2017
Beech Cottage Press
Paperback

ISBN: 978-09908330-4-8

Beech Cottage Press
Post Office Box 299
Lovingston, Virginia 22949
publishing@beechcottagepress.com

“Agriculture is our wisest pursuit, because it will in the end contribute most to real wealth, good morals and happiness.”

- Thomas Jefferson

“Be thou diligent to know the state of thy flocks, and look well to thy herds.”

- Proverbs 27:23

“I do not believe there ever was any life more attractive to a vigorous young fellow than life on a cattle ranch.”

- Theodore Roosevelt

CALVING LOG

Dam	Sire	Born	Sex	Weight	Tag

Weaned	Weight	Notes

Dam	Sire	Born	Sex	Weight	Tag

Weaned	Weight	Notes

Dam	Sire	Born	Sex	Weight	Tag

Weaned	Weight	Notes

Dam	Sire	Born	Sex	Weight	Tag

Weaned	Weight	Notes

Dam	Sire	Born	Sex	Weight	Tag

Weaned	Weight	Notes

Dam	Sire	Born	Sex	Weight	Tag

Weaned	Weight	Notes

HERD HEALTH

Animal	Date	Shots	Notes

Animal	Date	Shots	Notes

Animal	Date	Shots	Notes

Animal	Date	Shots	Notes

Animal	Date	Shots	Notes

Animal	Date	Shots	Notes

Animal	Date	Shots	Notes

Animal	Date	Shots	Notes

Date

/ /

Weather

On the Land

The Cows

Off the Land

Date

/ /

Weather

On the Land

The Cows

Off the Land

Date

/ /

Weather

On the Land

The Cows

Off the Land

Date

/ /

Weather

On the Land

The Cows

Off the Land

Date

/ /

Weather

On the Land

The Cows

Off the Land

Date

/ /

Weather

On the Land

The Cows

Off the Land

Date

/ /

Weather

On the Land

The Cows

Off the Land

Date

/ /

Weather

On the Land

The Cows

Off the Land

Date

/ /

Weather

On the Land

The Cows

Off the Land

Date

/ /

Weather

On the Land

The Cows

Off the Land

Date

/ /

Weather

On the Land

The Cows

Off the Land

Date

/ /

Weather

On the Land

The Cows

Off the Land

Date

/ /

Weather

On the Land

The Cows

Off the Land

Date

/ /

Weather

On the Land

The Cows

Off the Land

Date

/ /

Weather

On the Land

The Cows

Off the Land

Date

/ /

Weather

On the Land

The Cows

Off the Land

Date

/ /

Weather

On the Land

The Cows

Off the Land

Date

/ /

Weather

On the Land

The Cows

Off the Land

Date

/ /

Weather

On the Land

The Cows

Off the Land

Date

/ /

Weather

On the Land

The Cows

Off the Land

Date

/ /

Weather

On the Land

The Cows

Off the Land

Date

/ /

Weather

On the Land

The Cows

Off the Land

Date

/ /

Weather

On the Land

The Cows

Off the Land

Date

/ /

Weather

On the Land

The Cows

Off the Land

Date

/ /

Weather

On the Land

The Cows

Off the Land

Date

/ /

Weather

On the Land

The Cows

Off the Land

Date

/ /

Weather

On the Land

The Cows

Off the Land

Date

/ /

Weather

On the Land

The Cows

Off the Land

Date

/ /

Weather

On the Land

The Cows

Off the Land

Date

/ /

Weather

On the Land

The Cows

Off the Land

Date

/ /

Weather

On the Land

The Cows

Off the Land

Date

/ /

Weather

On the Land

The Cows

Off the Land

Date

/ /

Weather

On the Land

The Cows

Off the Land

Date

/ /

Weather

On the Land

The Cows

Off the Land

Date

/ /

Weather

On the Land

The Cows

Off the Land

Date

/ /

Weather

On the Land

The Cows

Off the Land

Date

/ /

Weather

On the Land

The Cows

Off the Land

Date

/ /

Weather

On the Land

The Cows

Off the Land

Date

/ /

Weather

On the Land

The Cows

Off the Land

Date

/ /

Weather

On the Land

The Cows

Off the Land

Date

/ /

Weather

On the Land

The Cows

Off the Land

Date

/ /

Weather

On the Land

The Cows

Off the Land

Date

/ /

Weather

On the Land

The Cows

Off the Land

Date

/ /

Weather

On the Land

The Cows

Off the Land

Date

/ /

Weather

On the Land

The Cows

Off the Land

Date

/ /

Weather

On the Land

The Cows

Off the Land

Date

/ /

Weather

On the Land

The Cows

Off the Land

Date

/ /

Weather

On the Land

The Cows

Off the Land

Date

/ /

Weather

On the Land

The Cows

Off the Land

Date

/ /

Weather

On the Land

The Cows

Off the Land

Date

/ /

Weather

On the Land

The Cows

Off the Land

Date

/ /

Weather

On the Land

The Cows

Off the Land

Date

/ /

Weather

On the Land

The Cows

Off the Land

Date

/ /

Weather

On the Land

The Cows

Off the Land

Date

/ /

Weather

On the Land

The Cows

Off the Land

Date

/ /

Weather

On the Land

The Cows

Off the Land

Date

/ /

Weather

On the Land

The Cows

Off the Land

Date

/ /

Weather

On the Land

The Cows

Off the Land

Date

/ /

Weather

On the Land

The Cows

Off the Land

Date

/ /

Weather

On the Land

The Cows

Off the Land

Date

/ /

Weather

On the Land

The Cows

Off the Land

Date

/ /

Weather

On the Land

The Cows

Off the Land

Date

/ /

Weather

On the Land

The Cows

Off the Land

Date

/ /

Weather

On the Land

The Cows

Off the Land

Date

/ /

Weather

On the Land

The Cows

Off the Land

Date

/ /

Weather

On the Land

The Cows

Off the Land

Date

/ /

Weather

On the Land

The Cows

Off the Land

Date

/ /

Weather

On the Land

The Cows

Off the Land

Date

/ /

Weather

On the Land

The Cows

Off the Land

Date

/ /

Weather

On the Land

The Cows

Off the Land

Date

/ /

Weather

On the Land

The Cows

Off the Land

Date

/ /

Weather

On the Land

The Cows

Off the Land

Date

/ /

Weather

On the Land

The Cows

Off the Land

Date

/ /

Weather

On the Land

The Cows

Off the Land

Date

/ /

Weather

On the Land

The Cows

Off the Land

Date

/ /

Weather

On the Land

The Cows

Off the Land

Date

/ /

Weather

On the Land

The Cows

Off the Land

Date

/ /

Weather

On the Land

The Cows

Off the Land

Date

/ /

Weather

On the Land

The Cows

Off the Land

Date

/ /

Weather

On the Land

The Cows

Off the Land

Date

/ /

Weather

On the Land

The Cows

Off the Land

Date

/ /

Weather

On the Land

The Cows

Off the Land

Date

/ /

Weather

On the Land

The Cows

Off the Land

Date

/ /

Weather

On the Land

The Cows

Off the Land

Date

/ /

Weather

On the Land

The Cows

Off the Land

Date

/ /

Weather

On the Land

The Cows

Off the Land

Date

/ /

Weather

On the Land

The Cows

Off the Land

Date

/ /

Weather

On the Land

The Cows

Off the Land

Date

/ /

Weather

On the Land

The Cows

Off the Land

Date

/ /

Weather

On the Land

The Cows

Off the Land

Date

/ /

Weather

On the Land

The Cows

Off the Land

Date

/ /

Weather

On the Land

The Cows

Off the Land

Date

/ /

Weather

On the Land

The Cows

Off the Land

Date

/ /

Weather

On the Land

The Cows

Off the Land

Date

/ /

Weather

On the Land

The Cows

Off the Land

Date

/ /

Weather

On the Land

The Cows

Off the Land

Date

/ /

Weather

On the Land

The Cows

Off the Land

Date

/ /

Weather

On the Land

The Cows

Off the Land

Date

/ /

Weather

On the Land

The Cows

Off the Land

Date

/ /

Weather

On the Land

The Cows

Off the Land

Date

/ /

Weather

On the Land

The Cows

Off the Land

Date

/ /

Weather

On the Land

The Cows

Off the Land

Date

/ /

Weather

On the Land

The Cows

Off the Land

Date

/ /

Weather

On the Land

The Cows

Off the Land

Date

/ /

Weather

On the Land

The Cows

Off the Land

Date

/ /

Weather

On the Land

The Cows

Off the Land

Date

/ /

Weather

On the Land

The Cows

Off the Land

Date

/ /

Weather

On the Land

The Cows

Off the Land

Date

/ /

Weather

On the Land

The Cows

Off the Land

Date

/ /

Weather

On the Land

The Cows

Off the Land

Date

/ /

Weather

On the Land

The Cows

Off the Land

Date

/ /

Weather

On the Land

The Cows

Off the Land

Date

/ /

Weather

On the Land

The Cows

Off the Land

Date

/ /

Weather

On the Land

The Cows

Off the Land

Date

/ /

Weather

On the Land

The Cows

Off the Land

Date

/ /

Weather

On the Land

The Cows

Off the Land

Date

/ /

Weather

On the Land

The Cows

Off the Land

Date

/ /

Weather

On the Land

The Cows

Off the Land

Date

/ /

Weather

On the Land

The Cows

Off the Land

Date

/ /

Weather

On the Land

The Cows

Off the Land

Date

/ /

Weather

On the Land

The Cows

Off the Land

Date

/ /

Weather

On the Land

The Cows

Off the Land

Date

/ /

Weather

On the Land

The Cows

Off the Land

Date

/ /

Weather

On the Land

The Cows

Off the Land

Date

/ /

Weather

On the Land

The Cows

Off the Land

Date

/ /

Weather

On the Land

The Cows

Off the Land

Date

/ /

Weather

On the Land

The Cows

Off the Land

Date

/ /

Weather

On the Land

The Cows

Off the Land

Date

/ /

Weather

On the Land

The Cows

Off the Land

Date

/ /

Weather

On the Land

The Cows

Off the Land

Date

/ /

Weather

On the Land

The Cows

Off the Land

Date

/ /

Weather

On the Land

The Cows

Off the Land

Date

/ /

Weather

On the Land

The Cows

Off the Land

Date

/ /

Weather

On the Land

The Cows

Off the Land

SALES		Date	
Animal	Weight	Price	Total

SALES		Date	
Animal	Weight	Price	Total

SALES		Date	
Animal	Weight	Price	Total

SALES		Date	
Animal	Weight	Price	Total

SALES		Date	
Animal	**Weight**	**Price**	**Total**

SALES		Date	
Animal	Weight	Price	Total

SALES		Date	
Animal	Weight	Price	Total

SALES		Date	
Animal	Weight	Price	Total

Made in the USA
Monee, IL
04 November 2026

67377182R00095